This book is dedicated to my very own creatures:
Andrea, **Erica** and **Sophia**.

It was tweaked and crafted by the inimitable
Chris Mills and the eagle-eyed **Janek Popiolek**.

Creature's Features

PB ISBN:978-0-9931730-0-4
A catalogue record for this book is available from the British Library.

Professor Hubert Merryweather

Now, sit up - pay attention, please!

To go hunting creatures is a difficult task.

'What equipment is needed?' most people ask.

You'll need crampons, a rucksack, magnifying glass,

Some sarnies, an apple, some soup in a flask.

An atlas is useful and also a map,

Of course... if you get lost you can always just ask

A native, or any tour guides that you pass:

'The way to the Jungle - can I get there by car?'

'Go left over there, but don't go too far.

When you're getting tired, turn right at the star,

But don't drive through the quicksand - you won't get very far.'

But above, all the things that you really must have

Are: brave heart, strong stomach, a waterproof pad

A camera is useful to prove what you've seen,

Or people won't believe that it wasn't a dream.

Now if this all sounds tough,

Or it's now time for lunch...

...don't worry I've done it, and gathered a bunch

Of the most bizarre creatures and all of their features.

So, turn o'er the page

Breathe deep, take a look,

At the curious beasties who lurk in this book.

Subject 001: The Spog

The Spog is the cutest of pond dwelling frogs.

It loves nothing more than relaxing on logs.

But the Spog is more dangerous than it might first appear,

This one here scared an elephant only last year.

You see...

The Spog has a venomous sting in its tail,

A poison so potent it could knock out a whale.

Now the Spog won't attack unless you provoke...

By taking a stick and giving him a poke,

Or you tell him he's ugly,

Or shout "you're so smelly!"

Or laugh really loud at the size of his belly.

So, unless you are mean, or perhaps you're a fly,

There's really no reason to fear this guy.

BRAZIL

Subject 002: Peruvian Marmorat

The Marmorat is incredibly tall,

It dwarfs buildings and lamposts and that is not all,

It can skip up a mountain and juggle a tree,

And even a giant can just reach its knee.

It's taller than me,

And it's taller than you,

But there's no need to worry

Because it lives in Peru.

The Marmorat eats nearly 10 tonnes a day!

It doesn't eat grass and it doesn't eat hay,

The Marmorat likes to eat only snow,

Quite why this might be, it seems nobody knows.

I cannot imagine how eating cold rain

Can help you grow big enough to wrestle a train.

But eat snow it does, in a ravenous way,

A snowdrift for breakfast and dinner each day.

And that's why in Peru you can never find snow.

And if you don't believe me, why don't you just go?

Subject 003: Bog Eyed Trout

The Bog Eyed Trout,

Has a very large mouth,

Which it uses to chew all its food.

It never sits down,

To eat its next meal,

Which some people might think a tad rude.

But it's not so surprising -

Fish don't just stop for dinner,

And the Bog Eyed Trout is a very strong swimmer.

So... with a drink in one fin,

And some cake in the other,

The trout munches while swimming around: no bother.

JAPAN

Subject 004: The Ravenous Cavernous Bat

Flippity flap is the sound of the bat
As it moves in the shade, of the cave where it made...
The sound of the flippity flap.

The sound of the bat is all that you'll know
If you head to the cave where the dreaded bats go...
Flippity flippity flap.

The echoes are heard by all who come here,
You'll see nothing but darkness, but hear in your ear...
The sound of the flippity flap.

So a torch would be useful except for the fact,
That the light will scare off any sign of the bat,
So the Ravenous Cavernous Bat can't be seen,
Except in this book, or a good magazine.

CANADA

Subject 005: The Chattering Turtle

The Chattering Turtle looks really quite scary,

Its very appearance can make people wary.

Its armour, its tail, its unicorn nose,

The spikes and the razor sharp claws on its toes.

But the truth is the Chattering Turtle is meek,

When threatened it cowers - sobbing and weak.

'Then how does it stop being eaten?' you cry.

'How does it stop being chattering pie?'

The answer you'd know if you'd heard the beast sing.

The shrieking, the wailing - the almighty din,

Is just the beginning of the pain that you're in,

The hideous noise will shatter your ears,

You'll be screaming in pain, with your eyes full of tears.

So take heed - if you meet one out in the wild,

Just smile and move on... be a clever child.

SOUTH AFRICA

Subject 006: Croc-A-Roc

It lies in the sun, tongue unfurled like a treat

Pink, ripe and tasty

And ready to eat...

Along comes a bird - unsuspecting, unthinking

The Croc-A-Roc waits - unmoving, unblinking.

When the bird puts its head

In the Croc-A-Roc's mouth

Snap!

Go the jaws...

...and the bird won't fly south.

The Croc-A-Roc's happy, its meal was incredible.

The bird is quite sad: it's just far too edible!

Now the only thought left inside Croc's little brain

Is what to eat next - the beast's hungry again...

NIGERIA

Subject 007: The Thunder Thighed Hen

The Thunder Thighed Hen makes an ideal pet,

They're pleasant and chirpy, they're friendly to Vets.

They only eat worms, so their upkeep is low

And if they escape, they can only run slow.

They can't really fly or jump over a fence

They wouldn't even try - as they are rather dense.

The only bad thing about these thunderous hens

Is: try as you might you can't really tame them.

I had one myself so I know this is true,

I tried everything you can possibly do.

I gave it a sugar lump; asked it to jump,

I tickled its chin and asked it to sing,

I bought it some string and asked it to juggle,

It acted quite bored, and got into a muddle.

I selected a stick so we could play fetch,

But it simply ignored me: the ignorant wretch.

So the hen's a great pet 'cause it's easy to keep
It has lovely soft feathers and a delightful beak,
But it won't perform tricks or even play dead,
Except if you whack it, real hard, in the head,
(Which I did, but only for research's sake).

Subject 008: The Hook-Nosed Ratypus

The Hook-Nosed Ratypus is a very strange thing,

It can't really walk and it can't really swim.

So it strolls down the river,

Swims down the path,

And wishes that it had been born a giraffe.

AUSTRALIA

Subject 009: The Kan-kan-go-ala

The Kan-kan-go-ala might look really cute,

Like it loves to sort flowers and eats only fruit.

But the kan-kan-go-ala enjoys nothing more,

Than tormenting poor boys and girls by the score...

Why only last week it clawed a boy's feet,

As it cackled and chased the poor soul down the street,

Then it pinched a poor girl who had done nothing wrong,

And pulled on her pigtails while singing a song.

But be glad: for the beast is really quite rare

And the chances of meeting one's gaze anywhere,

Are a million to one, or at least roughly there...

Abouts.

Subject 0010:

The scariest creature I ever did see?

It had two arms, two legs and looked rather like me.

These fearsome creatures with most bizarre features

Can be found in the day looked after by teachers.

Then you'll find them by night tormenting their dads,

Or scaring their mothers and making them sad.

They're tall short and skinny,

Some are boys, some are girls.

With big toothy smiles,

And freckles and curls.

They're pleasant and foul,

Bad tempered and nice,

Rude and polite,

Maybe foolish or wise.

They answer to 'Johnny, Kay, Ruby or Jack,

Delores or Maude, Shadrack and Pratt.'

They often look similar, but no two are the same

Apart from those twins... Richard and Zayne.

You'll find them stuck indoors, on cold-sad-rainy-days

Or frolicking round, when they're outside at play.

You know that I'm talking 'bout boys and 'bout girls

The type that you find all over the world.

They might not seem scary, but take it from me,

If you see a boy, or a girl: turn and flee!

www.ingramcontent.com/pod-product-compliance
Lightning Source LLC
Chambersburg PA
CBHW052046190326
41520CB00002BA/203